Collins

easy le

Fractions bumper book

Ages 5–7

Brad Thompson

How to use this book

- Easy Learning bumper books help your child improve basic skills, build confidence and develop a love of learning.

- Find a quiet, comfortable place to work, away from distractions.

- Get into a routine of completing one or two bumper book pages with your child each day.

- At the end of each double page, ask your child to circle the star that matches how many questions they have completed:

Some = half or fewer   Most = more than half   All = all of the questions

- The progress certificate at the back of this book will help you and your child keep track of how many  have been circled.

- Encourage your child to work through all of the questions eventually and praise them for completing the progress certificate.

Parent tip
Look out for tips on how to help your child learn.

- Ask your child to find and count the little mice that are hidden throughout this book.

- This will engage them with the pages of the book and get them interested in the activities.

(Don't count this one.)

Published by Collins
An imprint of HarperCollinsPublishers Ltd
1 London Bridge Street
London SE1 9GF

Browse the complete Collins catalogue at www.collins.co.uk

© HarperCollinsPublishers Ltd 2018

First published 2018

10 9 8 7 6 5 4 3 2 1

ISBN 9780008275488

The author asserts the moral right to be identified as the author of this work.

All rights reserved. No part of this publication may be reproduced, stored in a retrieval system, or transmitted, in any form or by any means, electronic, mechanical, photocopying, recording or otherwise, without the prior permission of Collins.

British Library Cataloguing in Publication Data.

A Catalogue record for this publication is available from the British Library.

All images and illustrations are © Shutterstock.com and © HarperCollinsPublishers

Author: Brad Thompson
Commissioning Editor: Michelle I'Anson
Project Manager: Rebecca Skinner
Cover Design: Sarah Duxbury
Text Design and Layout: QBS Learning
Production: Natalia Rebow
Printed and bound in China by RR Donnelley APS

Contents

Recognising one half $\left(\dfrac{1}{2}\right)$

1 Tick the object that is split into two equal parts.

2 Here is half an apple.

Tick the picture that shows the other half of the apple.

3 Tick the piece of paper that is cut in half.

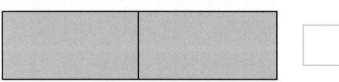

4 Tick the shape that is divided into halves.

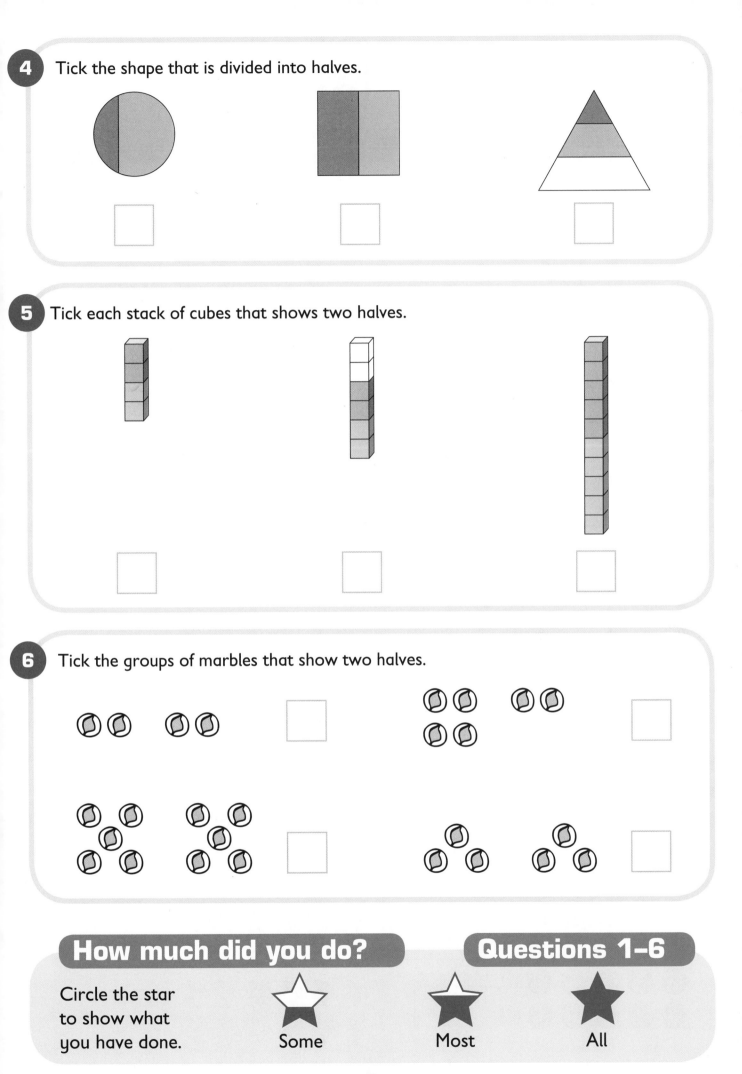

☐   ☐   ☐

5 Tick each stack of cubes that shows two halves.

☐   ☐   ☐

6 Tick the groups of marbles that show two halves.

☐   ☐

☐   ☐

How much did you do?   Questions 1–6

Circle the star to show what you have done.

Some   Most   All

Finding one half $\left(\dfrac{1}{2}\right)$

To find one half of something, you split it into two equal parts.

1 Colour half of each shape.

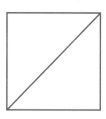

2 Using a pencil and a ruler, draw a straight line to divide each shape in half.

3 Find half of each amount.

 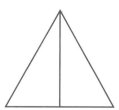

 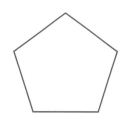

4 Complete each number sentence.

$\frac{1}{2}$ of 12 = ☐ $\frac{1}{2}$ of 4 = ☐ $\frac{1}{2}$ of 16 = ☐

$\frac{1}{2}$ of 8 = ☐ $\frac{1}{2}$ of 20 = ☐ $\frac{1}{2}$ of 2 = ☐

$\frac{1}{2}$ of 14 = ☐ $\frac{1}{2}$ of 6 = ☐ $\frac{1}{2}$ of 18 = ☐

5 All the numbers that go into the machine are halved.
Write the result for each number.

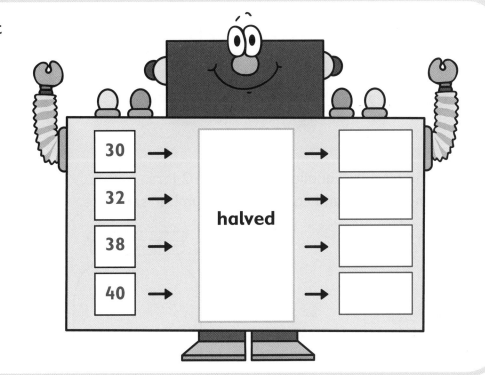

30 →   32 →   **halved**   38 →   40 →

6 Draw a line to match each pair of halves to the whole.

| 50 | 60 | 70 | 80 | 90 | 100 |

| 45 | 45 | 25 | 25 | 40 | 40 | 35 | 35 | 50 | 50 | 30 | 30 |

Sharing by two

When you share a set of objects equally between two, you halve the set.

1 Share the sweets equally between the 2 jars.
Draw the sweets and complete the number sentence.

12 shared by 2 = ☐

2 Share the biscuits equally between the 2 jars.
Draw the biscuits and complete the number sentence.

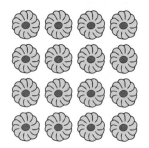

16 shared by 2 = ☐

3 Share the counters equally into 2 groups.
Draw the counters and complete the number sentence.

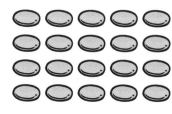

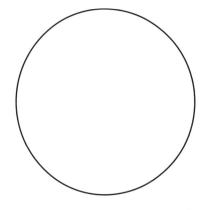
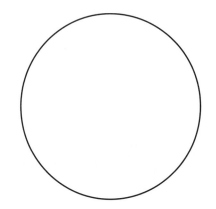

20 shared by 2 = ☐

4 Share the socks equally between the 2 laundry baskets.
Complete the number sentence.

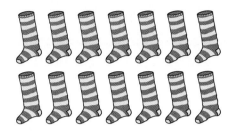

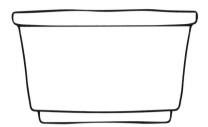

14 shared by 2 = [ ]

5 Share the flowers equally between the 2 vases.
Complete the number sentence.

10 shared by 2 = [ ]

6 Share the counters equally into 2 groups.
Complete the number sentence.

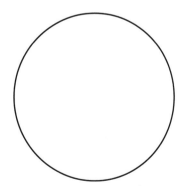

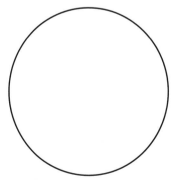

24 shared by 2 = [ ]

How much did you do?   Questions 1–6

Circle the star
to show what
you have done.

 Some   Most   All

Counting in halves

You can count forward in halves, e.g. $\frac{1}{2}$, 1, 1$\frac{1}{2}$, 2 ...

Each time you add $\frac{1}{2}$.

1 Fill in the box to complete the number line.
Complete the sentence below.

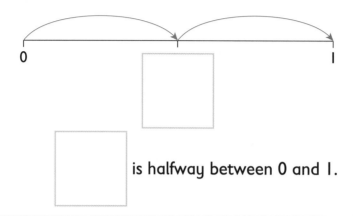

is halfway between 0 and 1.

2 Fill in the box to complete the number line.
Complete the sentence below.

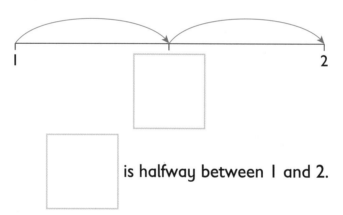

is halfway between 1 and 2.

3 Fill in the box to complete the number line.
Complete the sentence below.

Parent tip
Encourage your child to practise drawing and labelling their own number lines with whole numbers and halves.

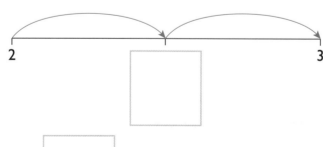

is halfway between 2 and 3.

4 Fill in the boxes to complete the number line.
Complete the sentence below.

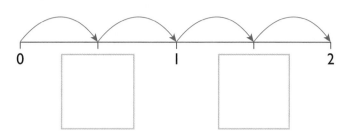

Between 0 and 2 there are ☐ jumps of one half.

5 Fill in the boxes to complete the number line.
Complete the sentence below.

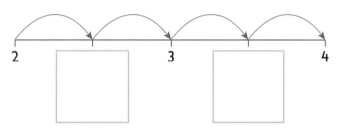

Between $2\frac{1}{2}$ and 4 there are ☐ jumps of one half.

6 Fill in the boxes to complete the number line.
Complete the sentence below.

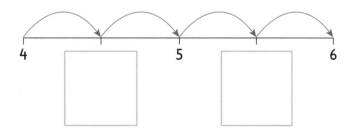

Between 4 and 6 there are ☐ jumps of one half.

Counting backward in halves

You can count backward in halves, e.g. $1\frac{1}{2}$, 1, $\frac{1}{2}$, 0.

Each time you take away $\frac{1}{2}$.

1 Start at 1 and count back in halves. Fill in the box to complete the number line. Complete the sentence below.

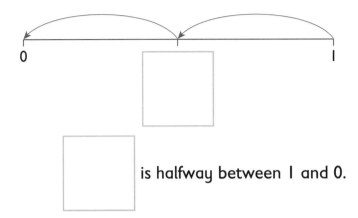

is halfway between 1 and 0.

2 Start at 2 and count back in halves. Fill in the box to complete the number line. Complete the sentence below.

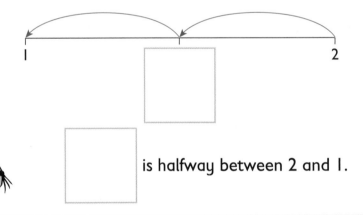

is halfway between 2 and 1.

3 Start at 3 and count back in halves. Fill in the box to complete the number line. Complete the sentence below.

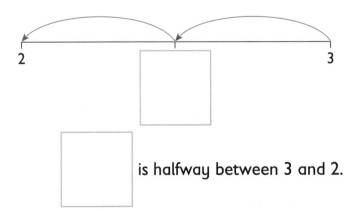

is halfway between 3 and 2.

Parent tip
Encourage your child to practise taking away fractions using food, e.g. cut some sandwiches into halves and ask your child to take away half at a time.

4 Start at 2 and count back in halves. Fill in the boxes to complete the number line. Complete the sentence below.

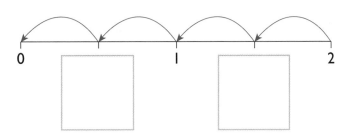

0   1   2

Between 2 and 0 there are ☐ jumps of $\frac{1}{2}$.

5 Start at 4 and count back in halves. Fill in the boxes to complete the number line. Complete the sentence below.

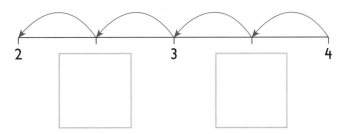

2   3   4

Between $3\frac{1}{2}$ and 2 there are ☐ jumps of $\frac{1}{2}$.

6 Start at 6 and count back in halves. Fill in the boxes to complete the number line. Complete the sentence below.

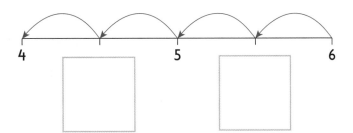

4   5   6

Between 6 and 4 there are ☐ jumps of $\frac{1}{2}$.

How much did you do?   Questions 1–6

Circle the star to show what you have done.

 Some

 Most

 All

Recognising one quarter $\left(\dfrac{1}{4}\right)$

1 Tick the shape that has one quarter shaded.

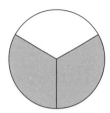

> **Parent tip**
> Read the title at the top of the page with your child. Point out that 'one quarter' can be written as $\dfrac{1}{4}$.

2 Tick each object that has one quarter missing.

3 Tick the box that shows where one quarter $\left(\dfrac{1}{4}\right)$ is on the number line.

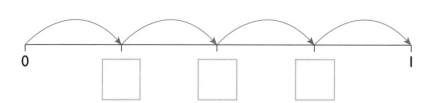

0   1

14

4 Tick each shape that is divided into quarters.

☐   ☐   ☐

5 Tick each array that shows quarters.
Then draw a ring around one quarter of the cubes in the array.

☐   ☐

☐

6 Tick the jar that has one quarter of the shells taken out.

☐   ☐

☐   ☐

How much did you do?   **Questions 1–6**

Circle the star
to show what
you have done.   ☆   ☆   ★
Some   Most   All

15

Finding one quarter $\left(\frac{1}{4}\right)$

1 Colour one quarter of each shape.

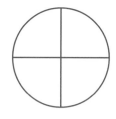 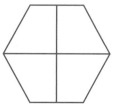 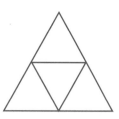

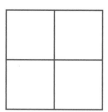  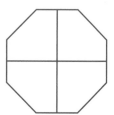

2 Colour one quarter of each set of fruit.

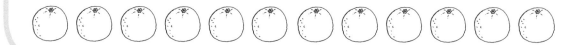

Parent tip
Point out that the 4 (the denominator) in $\frac{1}{4}$ tells you that the whole is divided into 4 equal parts.

3 Using a pencil and a ruler, draw two straight lines to divide each shape into quarters.

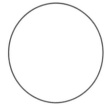 

 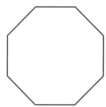

4 Colour one in every four parts of each shape.

When you finish, you will have coloured $\frac{1}{4}$ of the shape!

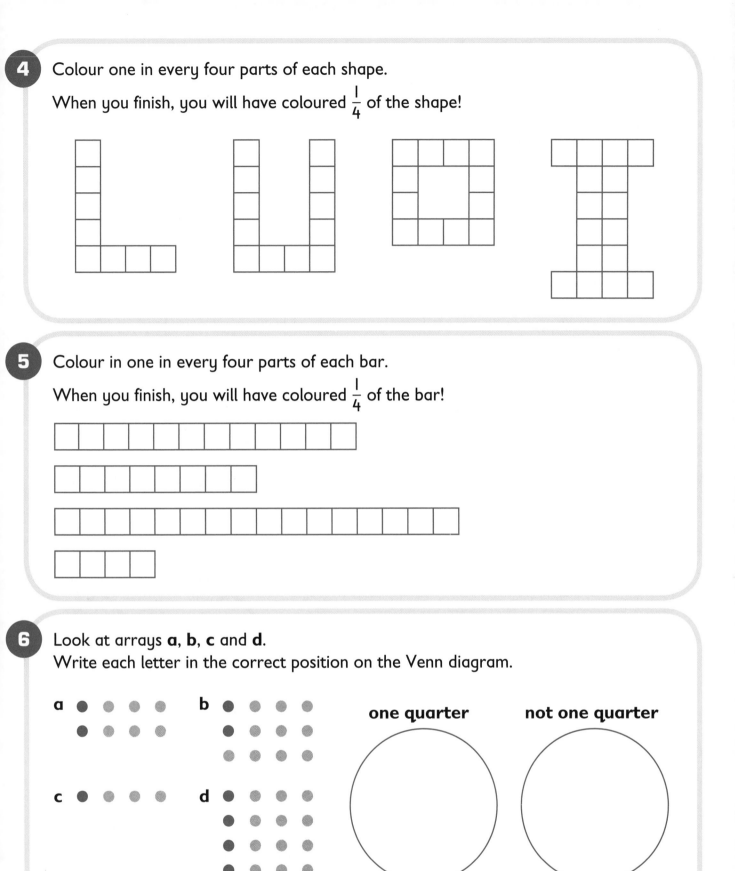

5 Colour in one in every four parts of each bar.

When you finish, you will have coloured $\frac{1}{4}$ of the bar!

6 Look at arrays **a**, **b**, **c** and **d**.
Write each letter in the correct position on the Venn diagram.

a

b

one quarter   not one quarter

c

d

17

Sharing by four

When you share a set of objects equally between four, you divide the set into quarters.

1 Share the crayons equally between the 4 trays.
Draw the crayons.
Write the number of crayons in each tray.

8 shared by 4 =

Parent tip
Ask your child to share 4, 8 or 12 objects into four equal groups.

2 Share the 16 letters equally between the 4 houses.
Draw the letters.
Write the number of letters that each house gets.

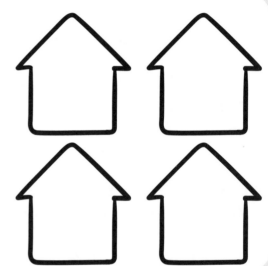

16 shared by 4 =

3 Share the gold coins equally between the 4 pirates.
Write the number of coins that each pirate gets.

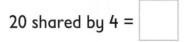

20 shared by 4 =

4 Share the buns equally between the 4 plates.
Write the number of buns on each plate.

12 shared by 4 =

5 Share the apples equally between the 4 horses.
Write the number of apples that each horse gets.

24 shared by 4 =

6 Share the tennis balls equally between the 4 players.
Write how many each player gets.

28 shared by 4 =

Recognising two quarters $\left(\dfrac{2}{4}\right)$

> Two quarters is two of four equal parts. This is the same as one half.

1 Tick each shape that has two quarters shaded.

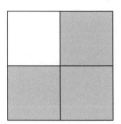 ☐

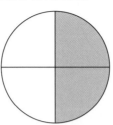 ☐

 ☐

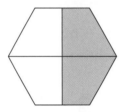 ☐

☐

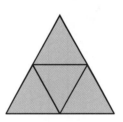 ☐

Parent tip
Cut a paper shape into halves and then into quarters to show your child that two quarters is the same as one half.

2 Draw a ring around each object with two quarters missing.

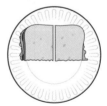

3 Tick the box that shows where two quarters $\left(\dfrac{2}{4}\right)$ is on the number line.

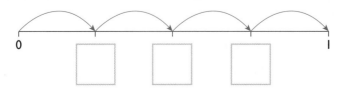

0   I

☐   ☐   ☐

4 Tick each shape that shows two quarters in the same colour.

☐ ☐ ☐

5 Tick each array that shows two quarters in the same colour.
Draw a ring around two quarters of the cubes in the array.

☐

☐

☐

6 Tick the picture that shows two quarters of the books in the box.

  ☐

  ☐

  ☐

Finding two quarters $\left(\dfrac{2}{4}\right)$

When finding two quarters, start by finding one quarter and then double it.

1 Colour two quarters of each shape.

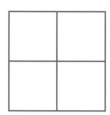 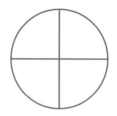 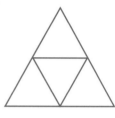

 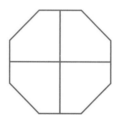 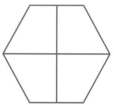

2 Colour two quarters of each set of fruit.

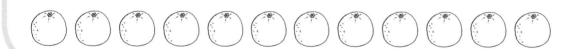

3 Look at the arrays.
Find two quarters
of each amount.
Write your answer.

  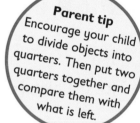

4 Colour two in every four parts of each shape.

When you finish, you will have coloured $\frac{2}{4}$ of the shape!

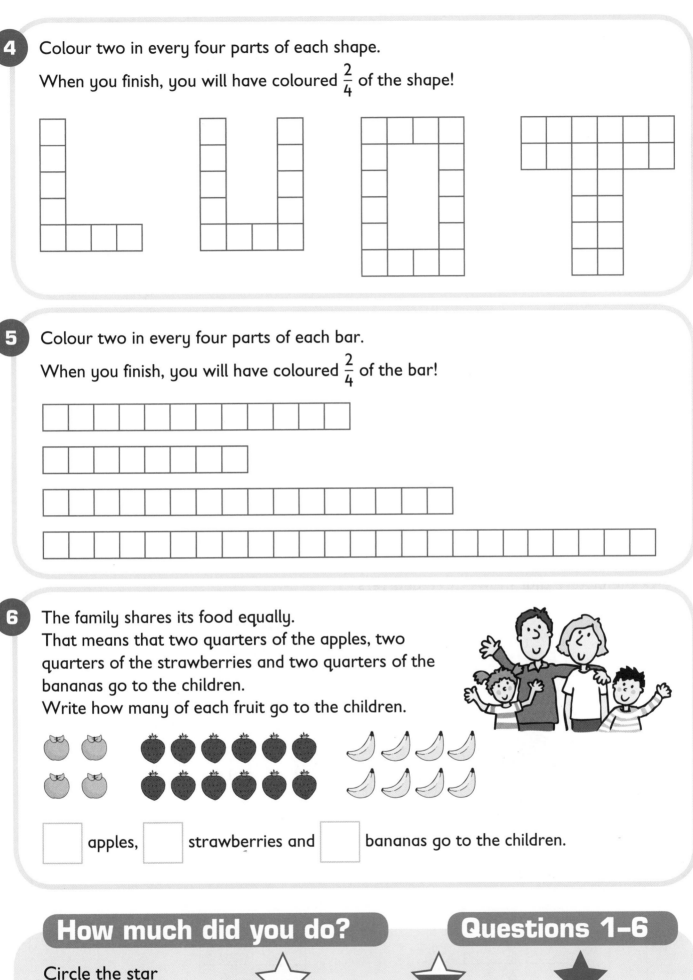

5 Colour two in every four parts of each bar.

When you finish, you will have coloured $\frac{2}{4}$ of the bar!

6 The family shares its food equally.
That means that two quarters of the apples, two quarters of the strawberries and two quarters of the bananas go to the children.
Write how many of each fruit go to the children.

☐ apples, ☐ strawberries and ☐ bananas go to the children.

Recognising three quarters $\left(\frac{3}{4}\right)$

Three quarters is bigger than one half, but less than one whole.

1 Tick each shape that has three quarters shaded.

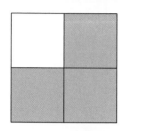

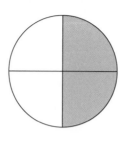

Parent tip
Cut a paper shape into quarters. Take one quarter away to show your child that three quarters is more than half, but less than a whole.

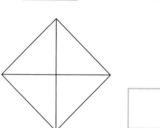

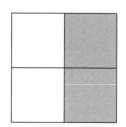

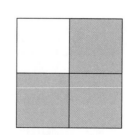

2 Tick the picture that shows three quarters.

3 Tick the shape that shows three quarters.

24

4 Tick the box if three quarters of the children are on the left-hand side of the seesaw.

5 Tick the box if three quarters of the balloons are blue.

6 Tick the washing line if three quarters of the T-shirts are white.

Finding three quarters $\left(\dfrac{3}{4}\right)$

When finding three quarters, start by finding one quarter and then add two more.

1 Colour three quarters of each shape.

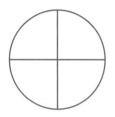 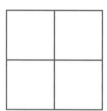 

 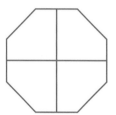 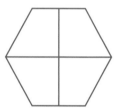

2 Colour three quarters of each set of fruit.

Parent tip
Encourage your child to divide a set of 4, 8 or 12 objects into four equal groups. Then put three of the groups together to find $\frac{3}{4}$.

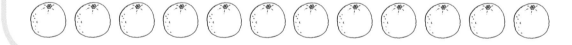

3 Look at the arrays.
Find three quarters of each amount.
Write your answer.

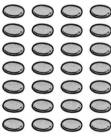

4 Colour three in every four parts of each shape.

When you finish, you will have coloured $\frac{3}{4}$ of the shape!

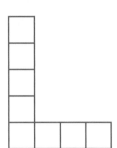 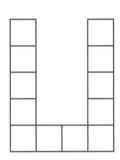 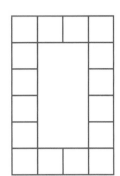 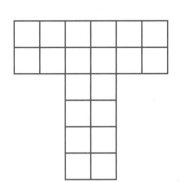

5 Colour three in every four parts of each bar.

When you finish, you will have coloured $\frac{3}{4}$ of the bar!

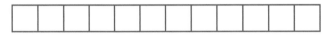

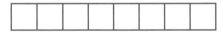

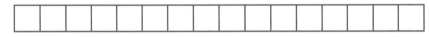

6 Complete the number sentences.

$\frac{3}{4}$ of 12 = ☐ $\frac{3}{4}$ of 4 = ☐ $\frac{3}{4}$ of 16 = ☐

$\frac{3}{4}$ of 8 = ☐ $\frac{3}{4}$ of 20 = ☐ $\frac{3}{4}$ of 24 = ☐

$\frac{3}{4}$ of 28 = ☐ $\frac{3}{4}$ of 40 = ☐ $\frac{3}{4}$ of 80 = ☐

How much did you do?   **Questions 1-6**

Circle the star to show what you have done.

 Some   Most   ★ All

Counting in quarters

You can count forward in quarters, e.g. $\frac{1}{4}$, $\frac{2}{4}$, $\frac{3}{4}$, **1, 1$\frac{1}{4}$ ...**
You add $\frac{1}{4}$ each time.

1 Fill in the boxes to complete the number line.
Complete the sentence below.

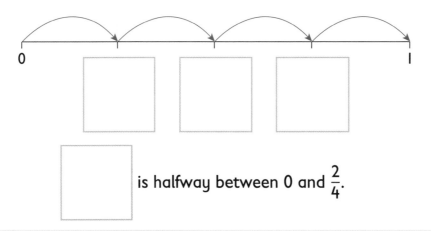

is halfway between 0 and $\frac{2}{4}$.

2 Fill in the boxes to complete the number line.
Complete the sentence below.

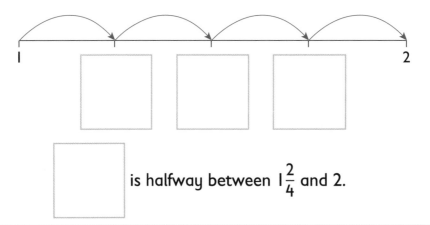

is halfway between 1$\frac{2}{4}$ and 2.

3 Fill in the boxes to complete the number line.
Complete the sentence below.

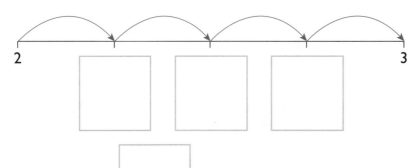

is halfway between 2 and 3.

Parent tip
Compare these number lines to the earlier ones with your child, so they can see that $\frac{2}{4}$ is the same as $\frac{1}{2}$.

4 Fill in the boxes to complete the number line.

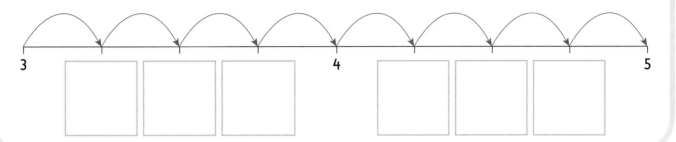

3   4   5

5 Fill in the boxes to complete the number line.

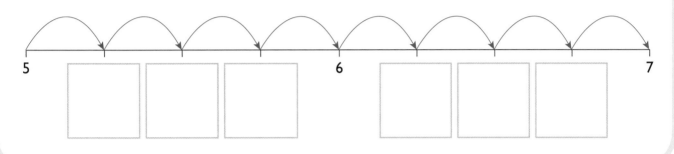

5   6   7

6 Fill in the boxes to complete the number lines.

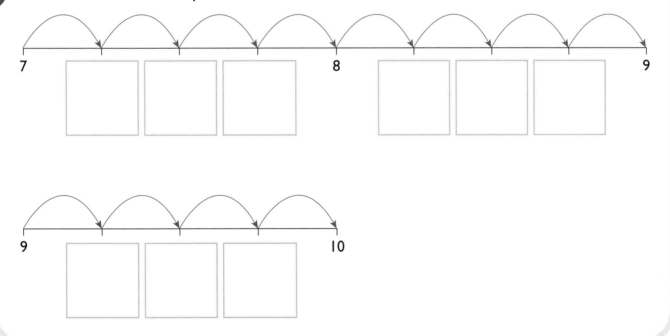

7   8   9

9   10

How much did you do?   Questions 1–6

Circle the star
to show what
you have done.

 Some   Most   All

Counting backward in quarters

You can count backward in quarters, e.g. $1\frac{1}{4}$, 1, $\frac{3}{4}$, $\frac{2}{4}$, $\frac{1}{4}$, 0.
You take away $\frac{1}{4}$ each time.

1 Start at 1 and count back in quarters. Fill in the boxes to complete the number line. Complete the sentence below.

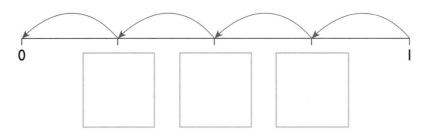

Between 1 and 0 there are ☐ jumps of $\frac{1}{4}$.

2 Start at 2 and count back in quarters. Fill in the boxes to complete the number line. Complete the sentence below.

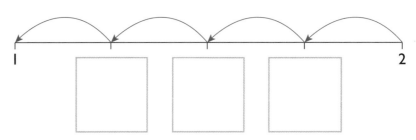

Between 2 and 1 there are ☐ jumps of $\frac{1}{4}$.

3 Start at 3 and count back in quarters. Fill in the boxes to complete the number line. Complete the sentence below.

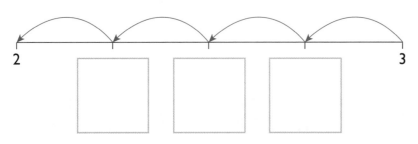

Parent tip
Ask your child to mark quarters on a piece of string and then cut the quarters off the end one at a time.

Between $2\frac{2}{4}$ and 2 there are ☐ jumps of $\frac{1}{4}$.

4 Start at 5 and count back in quarters. Fill in the boxes to complete the number line.

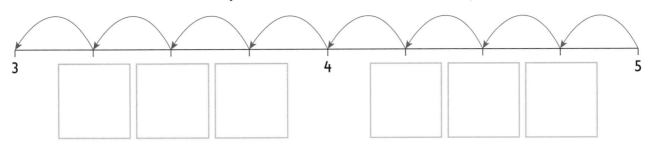

3   [ ] [ ] [ ]   4   [ ] [ ] [ ]   5

5 Start at 7 and count back in quarters. Fill in the boxes to complete the number line.

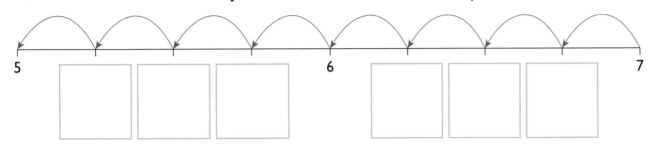

5   [ ] [ ] [ ]   6   [ ] [ ] [ ]   7

6 Start at 10 and count back in quarters. Fill in the boxes to complete the number lines.

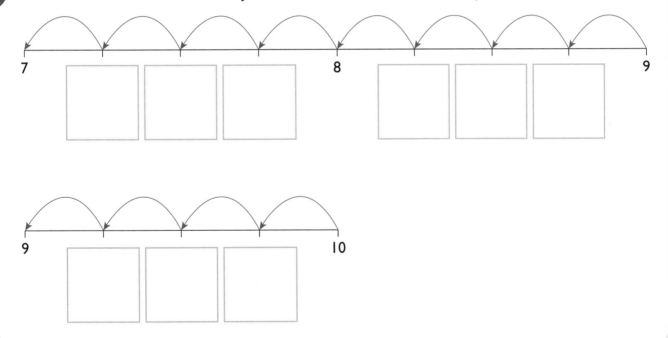

7   [ ] [ ] [ ]   8   [ ] [ ] [ ]   9

9   [ ] [ ] [ ]   10

Comparing halves and quarters

1 Write <, = or > between the fractions.

Parent tip
Cut some paper squares (that are the same size) into halves and quarters so that your child can physically compare the fractions.

2 Write <, = or > between the fractions.

3 Write <, = or > between the fractions.

4 Write <, = or > between the fractions.

5 Write <, = or > between the fractions.

 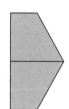

$\frac{3}{4}$ ☐ $\frac{1}{4}$

6 Write <, = or > between the fractions.

$\frac{1}{2}$ ☐ $\frac{2}{4}$

7 Write <, = or > between the fractions.

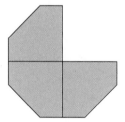 

$\frac{3}{4}$ ☐ $\frac{1}{2}$

Halves and quarters together

Halves and quarters can be added together to make a bigger fraction.

1 Here is quarter of a pizza and another quarter of a pizza.
What do you get when you add them together?

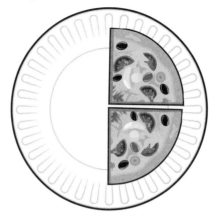

Parent tip
Cut a paper plate into quarters so your child can practise putting the quarters together in different ways.

2 Here is half a pizza and quarter of a pizza.
What do you get when you add them together?

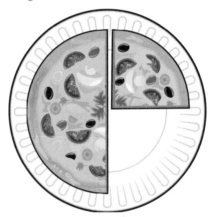

3 Here is quarter of a pizza, quarter of a pizza and another quarter of a pizza.
What do you get when you add them together?

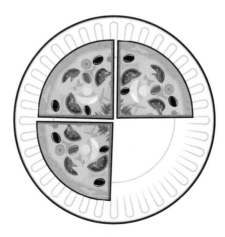

4 Colour in one more quarter of each shape.
Write the total fraction coloured.

5 Colour in one half.
Write the total fraction now coloured.

6 Colour in two quarters.
Write the total fraction now coloured.

Recognising one third $\left(\dfrac{1}{3}\right)$

A third is one of three equal parts.

1 Tick each object that is divided into three equal parts.

 ☐   ☐    ☐

 ☐   ☐    ☐

2 Tick the box that shows where one third $\left(\dfrac{1}{3}\right)$ is on the number line.

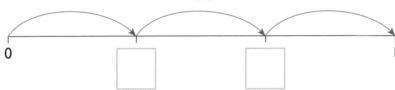

0   ☐   ☐   1

3 Tick each number that can be divided equally into thirds.

6 ☐   7 ☐

12 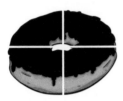 ☐   14 ☐

17  ☐   18 ☐

Parent tip
Ask your child to share 30 (and other multiples of 3) small objects (dried pasta works well) into groups of three.

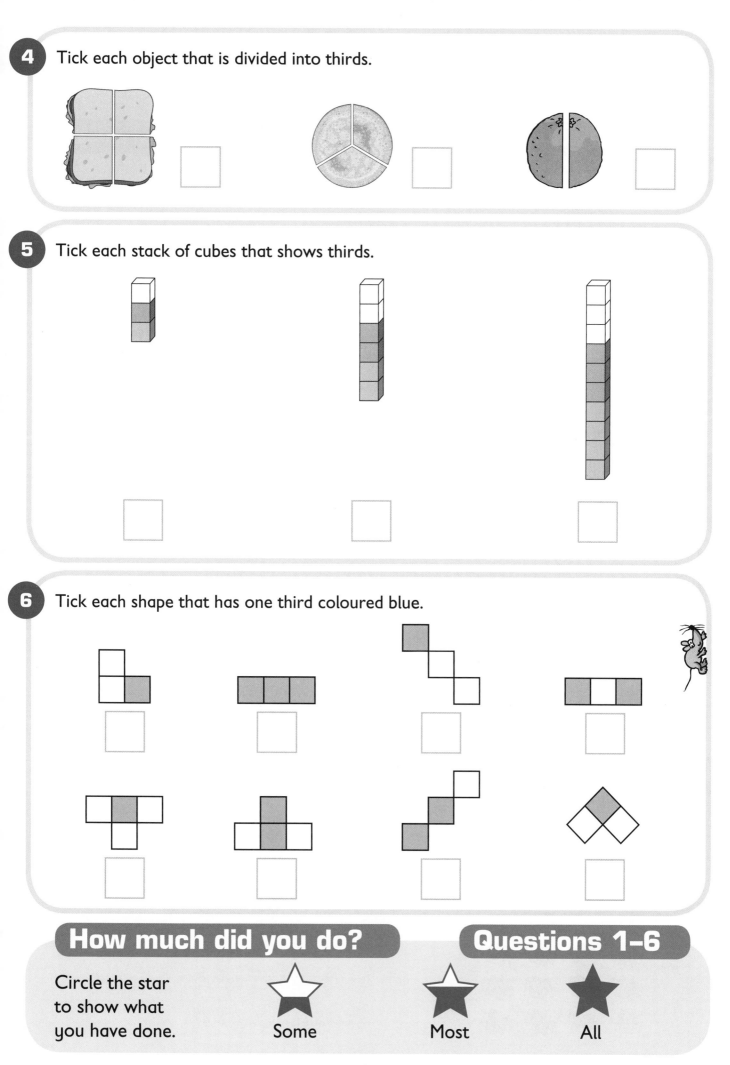

4 Tick each object that is divided into thirds.

5 Tick each stack of cubes that shows thirds.

6 Tick each shape that has one third coloured blue.

How much did you do?   **Questions 1–6**

Circle the star
to show what
you have done.

Some   Most   All

37

Finding one third $\left(\dfrac{1}{3}\right)$

To find one third of something, you divide it into three equal parts.

1 Colour one third of each shape.

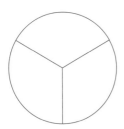
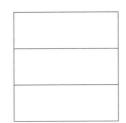
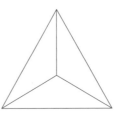

Parent tip
Draw shapes on a large piece of paper so your child can practise dividing them into thirds.

2 Using a pencil and a ruler, draw straight lines to divide each shape into thirds.

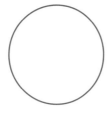

3 Find a third of each amount.
Write your answer in the box.

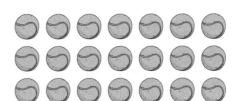

4 Colour one third of the flowers blue, one third red and one third yellow. Then complete the sentence.

$\frac{1}{3}$ of 9 = ☐

5 Colour one in every three parts of each bar.

When you finish, you will have coloured $\frac{1}{3}$ of the bar!

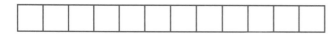

6 Complete the number sentences.

$\frac{1}{3}$ of 12 = ☐ $\frac{1}{3}$ of 15 = ☐ $\frac{1}{3}$ of 21 = ☐

$\frac{1}{3}$ of 3 = ☐ $\frac{1}{3}$ of 6 = ☐ $\frac{1}{3}$ of 18 = ☐

Sharing by three

1 Share the treats equally between the 3 dog bowls.
Draw the treats. Write how many treats each dog gets.

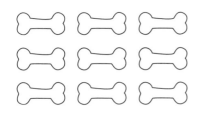  

9 shared by 3 = ☐

2 Share the beads equally between the 3 necklaces.
Draw the beads. Complete the number sentence.

18 shared by 3 = ☐

3 Share the counters equally into 3 groups.
Write how many counters there are in each group.

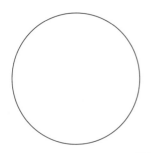 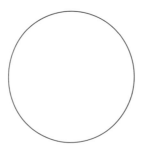 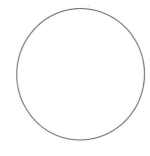

24 shared by 3 = ☐

Parent tip
Give your child sets of 3, 6, 9 or 12 objects and ask them to share them into three equal groups.

4 Share the pencils equally between the 3 pots.
Complete the number sentence.

12 shared by 3 = ☐

5 Share the cupcakes equally between the 3 plates.
Complete the number sentence.

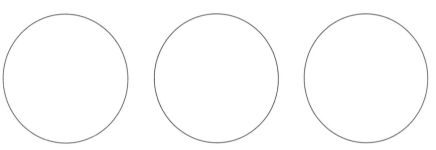

15 shared by 3 = ☐

6 Share the books equally between 3.
Complete the number sentence.

30 shared by 3 = ☐

Counting in thirds

You can count forward and backward in thirds, e.g. $\frac{1}{3}$, $\frac{2}{3}$, 1, $1\frac{1}{3}$ ...
You add or take away $\frac{1}{3}$ each time.

1 Fill in the boxes to complete the number line.
Complete the sentence below.

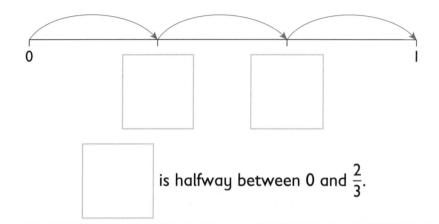

is halfway between 0 and $\frac{2}{3}$.

2 Fill in the boxes to complete the number line.
Complete the sentence below.

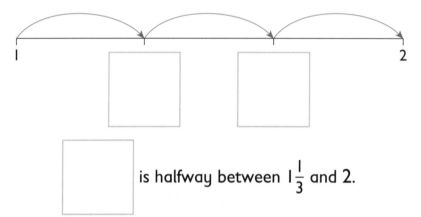

is halfway between $1\frac{1}{3}$ and 2.

3 Fill in the boxes to complete the number line.
Complete the sentence below.

Parent tip
Encourage your child to count up and down in thirds aloud.

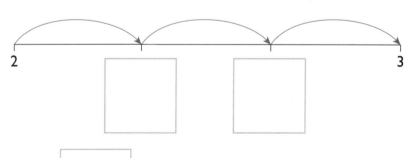

is halfway between 2 and $2\frac{2}{3}$.

4 Start at 1 and count back in thirds.
Fill in the boxes to complete the number line.
Complete the sentence below.

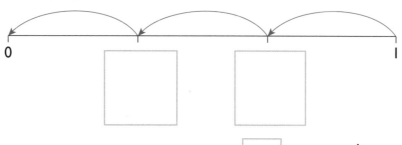

Between 1 and 0 there are ☐ jumps of $\frac{1}{3}$.

5 Start at 2 and count back in thirds.
Fill in the boxes to complete the number line.
Complete the sentence below.

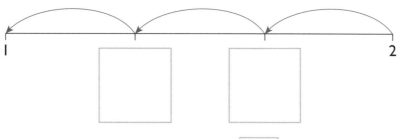

Between 2 and 1 there are ☐ jumps of $\frac{1}{3}$.

6 Start at 3 and count back in thirds.
Fill in the boxes to complete the number line.
Complete the sentence below.

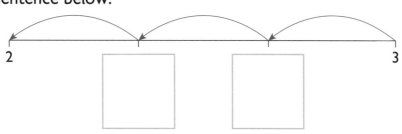

Between $2\frac{2}{3}$ and 2 there are ☐ jumps of $\frac{1}{3}$.

How much did you do?   Questions 1–6

Circle the star
to show what
you have done.

 Some   Most   All

Ordering fractions

1 Look at the fractions of pizza.
Write the fractions in order, from the smallest to the largest.

 $\frac{1}{2}$ $\frac{3}{4}$ $\frac{1}{4}$ 1

2 Look at the coloured parts of the shapes.
Write <, > or = in each box to compare the fractions shown.

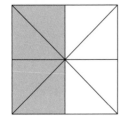 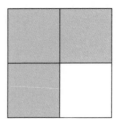   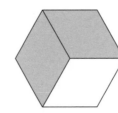 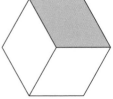

 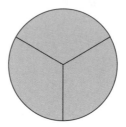

3 Write the missing fraction.

0        $\frac{1}{2}$ $\frac{3}{4}$ 1

4 Place the fractions in order from 0 (nothing) to 1 (one whole) on the number line.

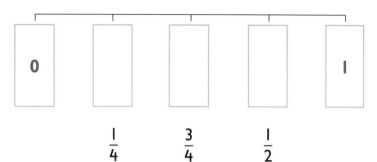

$\dfrac{1}{4}$ $\dfrac{3}{4}$ $\dfrac{1}{2}$

5 Write <, > or = in each box to compare the amounts shown by blue dots in the arrays.

$\dfrac{1}{2}$ of 6 ☐ $\dfrac{1}{2}$ of 8 $\dfrac{1}{2}$ of 14 ☐ $\dfrac{1}{2}$ of 10

$\dfrac{1}{4}$ of 8 ☐ $\dfrac{1}{4}$ of 12 $\dfrac{1}{4}$ of 20 ☐ $\dfrac{1}{4}$ of 16

6 Write these fractions in order, from the smallest to the largest.

$\dfrac{1}{3}$ $\dfrac{1}{4}$ $\dfrac{1}{2}$ $\dfrac{4}{4}$

Writing fractions

1 Write one quarter as a number.

2 Write one half as a number.

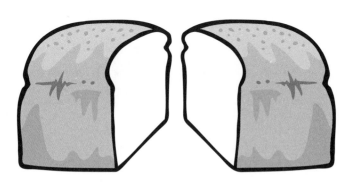

3 Write three quarters as a number.

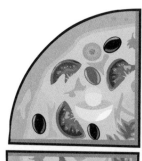 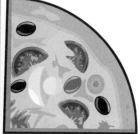

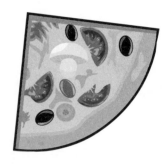

4 Write two quarters as a number.

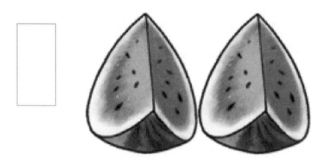

5 Write one third as a number.

6 Write each fraction in words.
Use the words in blue to help.

one   two   three   quarter   half   third

$\frac{1}{4}$ _____

$\frac{1}{2}$ _____

$\frac{2}{4}$ _____

$\frac{3}{4}$ _____

$\frac{1}{3}$ _____

Equivalent fractions

The same fraction can often be shown or written in different ways. Amounts that look different but are actually the same are 'equivalent'.

1 Write the fraction shown by the coloured part of each shape.

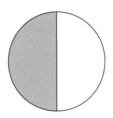   =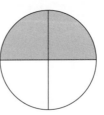

=

2 Write the fraction shown by the coloured part of each shape.

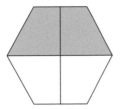   =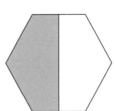

=

3 Write the fraction shown by the coloured part of each shape.

   =

=

4 Look at the first shape.
Colour the second shape to show
a fraction equivalent to the first.
Write the fraction under each shape.

$\square$ = $\square$

5 Look at the first shape.
Colour the second shape to show
a fraction equivalent to the first.
Write the fraction under each shape.

$\square$ = $\square$

6 Look at the first shape.
Colour the second shape to show
a fraction equivalent to the first.
Write the fraction under each shape.

Parent tip
Draw a large rectangle
and divide it into three
equal rows. Ask your
child to divide each row
into halves, quarters
and thirds and to label
each 'brick' to create
a fraction wall.

$\square$ = $\square$

How much did you do?   Questions 1–6

Circle the star
to show what
you have done.   Some   Most   All

49

Fractions of length

You can use your knowledge of fractions to find fractions of a length.

1 Where has the length of string been cut?
Tick the correct option.

Halfway along ☐

One quarter of the way along ☐

2 How far along the washing line is the T-shirt?
Tick the correct option.

Halfway along ☐

One quarter of the way along ☐

3 Where has the length of wood been cut?
Tick the correct option.

Halfway along ☐

One quarter of the way along ☐

One third of the way along ☐

Parent tip
Encourage your child to measure the lengths of different objects and then find one half, one quarter and one third of that length.

4 Find one quarter of each length of string.

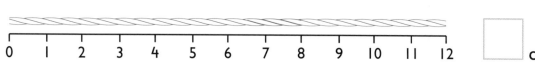

0   2   4   6   8

 cm

0   1   2   3   4

[ ] cm

0   1   2   3   4   5   6   7   8   9   10   11   12

[ ] cm

5 Find one half of each length of string.

0   3   6   9   12

[ ] cm

0   1   2   3   4   5   6   7   8   9   10

[ ] cm

6 Find three quarters of each length of string.

0   3   6   9   12

[ ] cm

0   2   4   6   8

[ ] cm

Fractions of mass

You can use your knowledge of fractions and apply it to weight or mass.

1 Look at the balance scales.
Find half of the total mass.
Complete the sentence below.

$\frac{1}{2}$ of 8 kg is ⬚ kg.

2 Look at the balance scales.
Find one quarter of the total mass.
Complete the sentence below.

$\frac{1}{4}$ of 8 kg is ⬚ kg.

3 Look at the balance scales.
Find three quarters of the total mass.
Complete the sentence below.

Parent tip
Get your child to weigh out a quantity of dry pasta (or similar) and then find different fractions of that amount.

$\frac{3}{4}$ of 16 kg is ⬚ kg.

4 Look at the weighing scales.
Find one quarter of each mass.

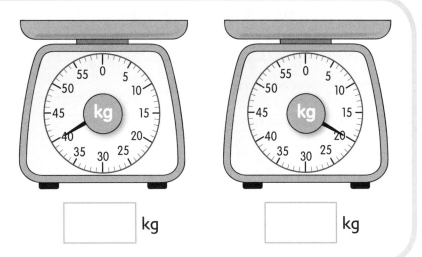

kg

kg

5 Look at the weighing scales.
Find one half of each mass.

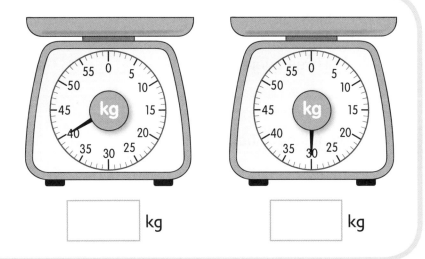

kg

kg

6 Look at the weighing scales.
Find three quarters of each mass.

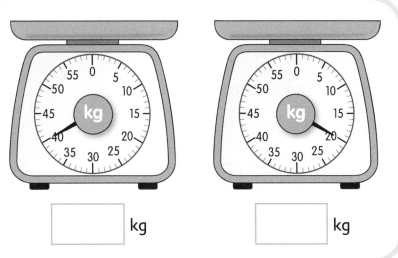

kg

kg

Fractions of capacity

1 Tick how full each measuring jug is.

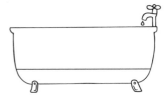 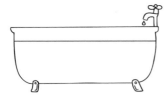 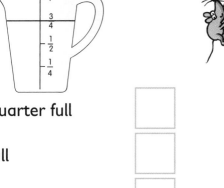

One quarter full ☐   One quarter full ☐

Half full ☐   Half full ☐

Three quarters full ☐   Three quarters full ☐

2 Tick how full each bath tub is.

One quarter full ☐   One quarter full ☐

Half full ☐   Half full ☐

Three quarters full ☐   Three quarters full ☐

3 Tick how full each drinking glass is.

Parent tip
Encourage your child to fill containers and describe how full they are using fractions vocabulary.

One quarter full ☐   One quarter full ☐

Half full ☐   Half full ☐

Three quarters full ☐   Three quarters full ☐

4 Find one quarter of the capacity of each jug.

☐ litre

☐ litres

5 Find one half of the capacity of each jug.

☐ litres

☐ litres

6 Find three quarters of the capacity of each jug.

☐ litres

☐ litres

How much did you do?   ## Questions 1–6

Circle the star
to show what
you have done.

 Some

 Most

 All

Word problems

1 Fatima invited 20 friends to her birthday party.
Half of the friends invited were girls and half were boys.

How many boys were invited to the party?

[ ] boys

2 Circle $\frac{2}{4}$ of the stamps.

Complete the sentence below.

$\frac{2}{4}$ of 8 stamps is [ ] stamps.

3 Peter had 15 strawberries.

He ate $\frac{1}{3}$ of them.

Parent tip
Encourage your child to use objects or produce a drawing of the problem. Ask them to explain their reasoning to you.

How many strawberries did he have left?

[ ] strawberries

4 Karim and his brother are sharing a pizza.
Their mum gives them half each.
Karim says,

'His half is bigger than mine!'

Is Karim correct?　　　yes ☐　　no ☐

Explain your answer.

_____

_____

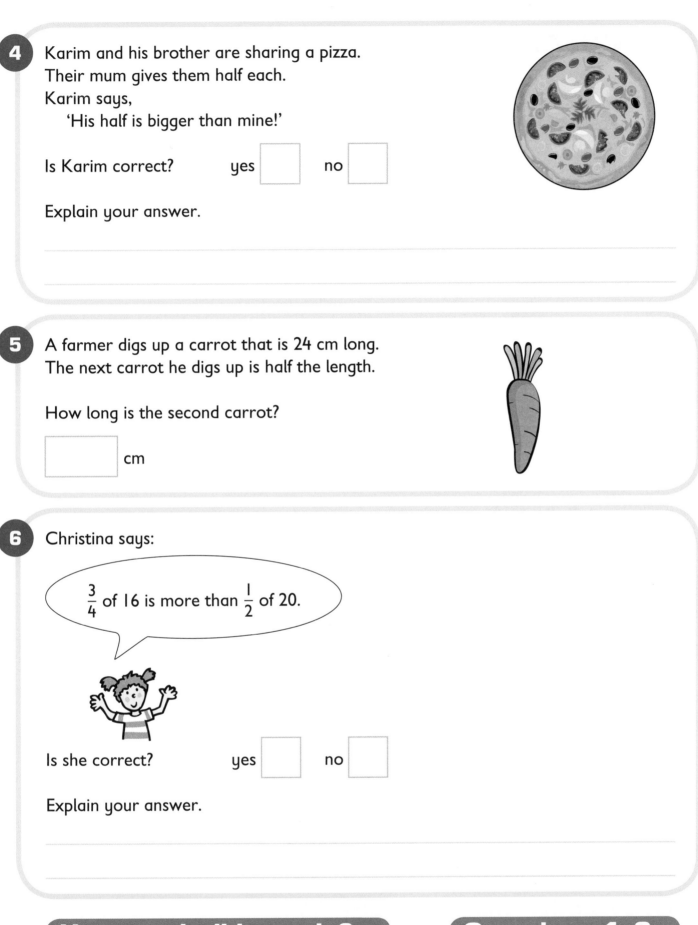

5 A farmer digs up a carrot that is 24 cm long.
The next carrot he digs up is half the length.

How long is the second carrot?

☐ cm

6 Christina says:

$\frac{3}{4}$ of 16 is more than $\frac{1}{2}$ of 20.

Is she correct?　　　yes ☐　　no ☐

Explain your answer.

_____

_____

How much did you do?　　　Questions 1–6

Circle the star
to show what
you have done.　　☆ Some　　★ Most　　★ All

57

Answers

Recognising one half $\left(\frac{1}{2}\right)$

Page 4

1

2

3

Page 5

4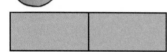

5

6

Finding one half $\left(\frac{1}{2}\right)$

Page 6

1 Either half may be coloured, e.g.

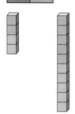

2 All shapes divided in half, e.g.

3 8, 6, 10, 12

Page 7

4 6, 2, 8, 4, 10, 1, 7, 3, 9

5 15, 16, 19, 20

6 50 = 25 + 25, 60 = 30 + 30, 70 = 35 + 35, 80 = 40 + 40, 90 = 45 + 45, 100 = 50 + 50

Sharing by two

Page 8

1 6   2 8   3 10

Page 9

4 7   5 5   6 12

Counting in halves

Page 10

1 $\frac{1}{2}$, $\frac{1}{2}$

2 $1\frac{1}{2}$, $1\frac{1}{2}$

3 $2\frac{1}{2}$, $2\frac{1}{2}$

Page 11

4 $\frac{1}{2}$, $1\frac{1}{2}$, 4

5 $2\frac{1}{2}$, $3\frac{1}{2}$, 3

6 $4\frac{1}{2}$, $5\frac{1}{2}$, 4

Counting backward in halves

Page 12

1 $\frac{1}{2}$, $\frac{1}{2}$

2 $1\frac{1}{2}$, $1\frac{1}{2}$

3 $2\frac{1}{2}$, $2\frac{1}{2}$

Page 13

4 $\frac{1}{2}$, $1\frac{1}{2}$, 4

5 $2\frac{1}{2}$, $3\frac{1}{2}$, 3

6 $4\frac{1}{2}$, $5\frac{1}{2}$, 4

Recognising one quarter $\left(\dfrac{1}{4}\right)$

Page 14

1

2

3
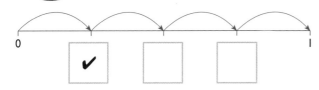

Page 15

4

5 Any 2 and 4 cubes may be ringed, e.g.

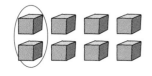
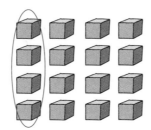

6

Finding one quarter $\left(\dfrac{1}{4}\right)$

Page 16

1 Any one quarter may be coloured, e.g.

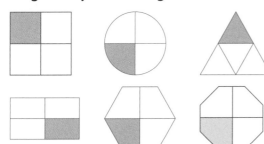

2 1 apple, 2 bananas and 3 oranges should be coloured.

3 All shapes divided into quarters, e.g.

Page 17

4 Any 2 parts shaded, 3 parts shaded, 3 parts shaded, 4 parts shaded, e.g.

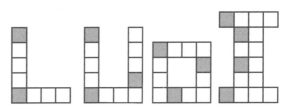

5 Any 3 parts shaded, 2 parts shaded, 4 parts shaded, 1 part shaded, e.g.

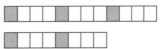

6
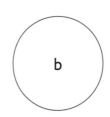

Sharing by four

Page 18

1 2   **2** 4   **3** 5

Page 19

4 3   **5** 6   **6** 7

Recognising two quarters $\left(\dfrac{2}{4}\right)$

Page 20

1

2

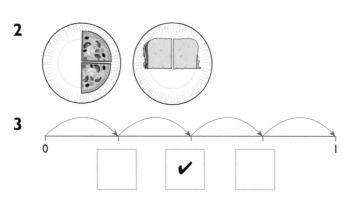

3

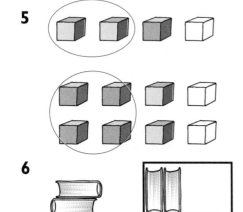

Page 21

4

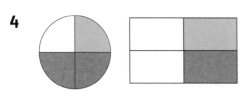

5

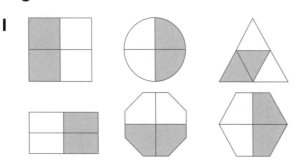

6

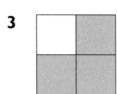

Finding two quarters $\left(\frac{2}{4}\right)$

Page 22

1

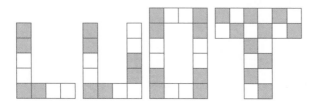

2   2 apples, 4 bananas and 6 oranges coloured.

3   2, 10, 6, 12

Page 23

4   Any 4 parts shaded, 6 parts shaded, 8 parts shaded, 10 parts shaded, e.g.

5   Any 6 parts shaded, 4 parts shaded, 8 parts shaded, 12 parts shaded, e.g.

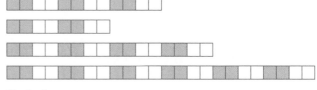

6   2, 6, 4

Recognising three quarters $\left(\frac{3}{4}\right)$

Page 24

1

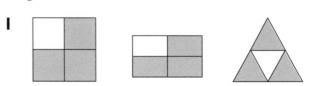

2

3

Page 25

4   Picture ticked.

5   Picture ticked.

6   Picture showing two blue T-shirts ticked.

Finding three quarters $\left(\frac{3}{4}\right)$

Page 26

1

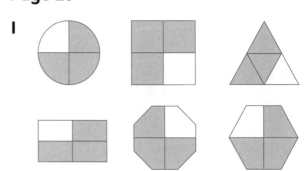

2   3 apples, 6 bananas and 9 oranges coloured.

3   3, 12, 9, 21

Page 27

4 Any 6 parts shaded, 9 parts shaded, 12 parts shaded, 15 parts shaded, e.g.

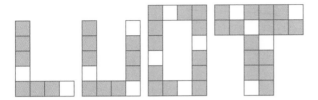

5 Any 9 parts shaded, 6 parts shaded, 12 parts shaded, 18 parts shaded, e.g.

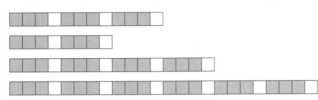

6 9, 3, 12, 6, 15, 18, 21, 30, 60

Counting in quarters

Page 28

For all questions, accept $\frac{1}{2}$ instead of $\frac{2}{4}$.

1 $\frac{1}{4}, \frac{2}{4}, \frac{3}{4}, 1\frac{1}{4}$

2 $1\frac{1}{4}, 1\frac{2}{4}, 1\frac{3}{4}, 1\frac{3}{4}$

3 $2\frac{1}{4}, 2\frac{2}{4}, 2\frac{3}{4}, 2\frac{2}{4}$

Page 29

4 $3\frac{1}{4}, 3\frac{2}{4}, 3\frac{3}{4}, 4\frac{1}{4}, 4\frac{2}{4}, 4\frac{3}{4}$

5 $5\frac{1}{4}, 5\frac{2}{4}, 5\frac{3}{4}, 6\frac{1}{4}, 6\frac{2}{4}, 6\frac{3}{4}$

6 $7\frac{1}{4}, 7\frac{2}{4}, 7\frac{3}{4}, 8\frac{1}{4}, 8\frac{2}{4}, 8\frac{3}{4}, 9\frac{1}{4}, 9\frac{2}{4}, 9\frac{3}{4}$

Counting backward in quarters

Page 30

For all questions, accept $\frac{1}{2}$ instead of $\frac{2}{4}$.

1 $\frac{1}{4}, \frac{2}{4}, \frac{3}{4}, 4$

2 $1\frac{1}{4}, 1\frac{2}{4}, 1\frac{3}{4}, 4$

3 $2\frac{1}{4}, 2\frac{2}{4}, 2\frac{3}{4}, 2$

Page 31

4 $3\frac{1}{4}, 3\frac{2}{4}, 3\frac{3}{4}, 4\frac{1}{4}, 4\frac{2}{4}, 4\frac{3}{4}$

5 $5\frac{1}{4}, 5\frac{2}{4}, 5\frac{3}{4}, 6\frac{1}{4}, 6\frac{2}{4}, 6\frac{3}{4}$

6 $7\frac{1}{4}, 7\frac{2}{4}, 7\frac{3}{4}, 8\frac{1}{4}, 8\frac{2}{4}, 8\frac{3}{4}, 9\frac{1}{4}, 9\frac{2}{4}, 9\frac{3}{4}$

Comparing halves and quarters

Page 32

1 >   **2** >   **3** <   **4** =

Page 33

5 >   **6** =   **7** >

Halves and quarters together

Page 34

1 $\frac{2}{4}$ or $\frac{1}{2}$ **2** $\frac{3}{4}$ **3** $\frac{3}{4}$

Page 35

4 $\frac{2}{4}$ or $\frac{1}{2}$, $\frac{3}{4}$, $\frac{2}{4}$ or $\frac{1}{2}$, $\frac{4}{4}$ or $\frac{2}{2}$ or 1

5 $\frac{3}{4}$, $\frac{4}{4}$ or 1, $\frac{2}{4}$ or $\frac{1}{2}$

6 $\frac{3}{4}$, $\frac{4}{4}$ or $\frac{2}{2}$ or 1, $\frac{3}{4}$, $\frac{2}{4}$ or $\frac{1}{2}$

Recognising one third $\left(\frac{1}{3}\right)$

Page 36

1

2

3 6, 12 and 18 ticked

Page 37

4

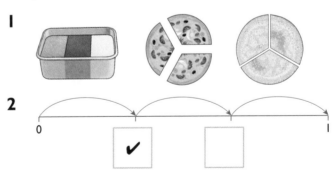

5

6

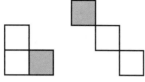

Finding one third $\left(\dfrac{1}{3}\right)$

Page 38

1 Any one third may be coloured, e.g.

2 All shapes divided into thirds, e.g.

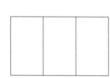

3   5, 7, 4, 6

Page 39

4   3

5 Any 4 parts shaded, 2 parts shaded, 5 parts shaded, 3 parts shaded, and 1 part shaded e.g.

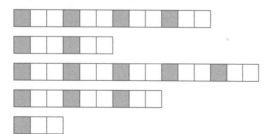

6   4, 5, 7, 1, 2, 6

Sharing by three

Page 40

1   3   2   6   3   8

Page 41

4   4   5   5   6   10

Counting in thirds

Page 42

1   $\dfrac{1}{3}, \dfrac{2}{3}, \dfrac{1}{3}$

2   $1\dfrac{1}{3}, 1\dfrac{2}{3}, 1\dfrac{2}{3}$

3   $2\dfrac{1}{3}, 2\dfrac{2}{3}, 2\dfrac{1}{3}$

Page 43

4   $\dfrac{1}{3}, \dfrac{2}{3}, 3$

5   $1\dfrac{1}{3}, 1\dfrac{2}{3}, 3$

6   $2\dfrac{1}{3}, 2\dfrac{2}{3}, 2$

Ordering fractions

Page 44

1   $\dfrac{1}{4}, \dfrac{1}{2}, \dfrac{3}{4}, 1$

2   <, >, =

3   $\dfrac{1}{4}$

Page 45

4 | 0 | $\dfrac{1}{4}$ | $\dfrac{1}{2}$ | $\dfrac{3}{4}$ | 1 |

5   <, >, <, >

6   $\dfrac{1}{4}, \dfrac{1}{3}, \dfrac{1}{2}, \dfrac{4}{4}$

Writing fractions

Page 46

1   $\dfrac{1}{4}$

2 $\frac{1}{2}$

3 $\frac{3}{4}$

Page 47

4 $\frac{2}{4}$

5 $\frac{1}{3}$

6 one quarter, one half, two quarters, three quarters, one third

Equivalent fractions

Page 48

1 $\frac{1}{2} = \frac{2}{4}$

2 $\frac{2}{4} = \frac{1}{2}$

3 $\frac{4}{4} = \frac{2}{2}$

Page 49

4 Any two quarters of the second square coloured, $\frac{1}{2} = \frac{2}{4}$

5 Any half of the second square coloured, $\frac{2}{4} = \frac{1}{2}$

6 Any two quarters of the second square coloured, $\frac{2}{4} = \frac{2}{4}$

Fractions of length

Page 50

1 Halfway along

2 One quarter of the way along

3 One third of the way along

Page 51

4 2 cm, 1 cm, 3 cm

5 6 cm, 5 cm

6 9 cm, 6 cm

Fractions of mass

Page 52

1 4 kg

2 2 kg

3 12 kg

Page 53

4 10 kg, 5 kg

5 20 kg, 15 kg

6 30 kg, 15 kg

Fractions of capacity

Page 54

1 Half full, Three quarters full

2 One quarter full, Three quarters full

3 Half full, One quarter full

Page 55

4 1 litre, 2 litres

5 5 litres, 10 litres

6 3 litres, 6 litres

Word problems

Page 56

1 10 boys

2 4 stamps circled, 4

3 10 strawberries

Page 57

4 no

One half is equal to one half. They are not halves if one slice is bigger than the other.

5 12 cm

6 yes

$\frac{3}{4}$ of 16 = 12 and $\frac{1}{2}$ of 20 = 10.

Check your progress

- Shade in the stars on the progress certificate to show how much you did. Shade one star for every ⭐ you circled in this book.
- If you have shaded fewer than 20 stars go back to the pages where you circled Some ☆ or Most ⭐ and try those pages again.
- If you have shaded 20 or more stars, well done!

Fractions Ages 5–7
Progress certificate

name _____   date _____

pages 4–5	pages 6–7	pages 8–9	pages 10–11	pages 12–13	pages 14–15	pages 16–17	pages 18–19	pages 20–21
☆ 1	☆ 2	☆ 3	☆ 4	☆ 5	☆ 6	☆ 7	☆ 8	☆ 9

pages 22–23	pages 24–25	pages 26–27	pages 28–29	pages 30–31	pages 32–33	pages 34–35	pages 36–37	pages 38–39
☆ 10	☆ 11	☆ 12	☆ 13	☆ 14	☆ 15	☆ 16	☆ 17	☆ 18

pages 40–41	pages 42–43	pages 44–45	pages 46–47	pages 48–49	pages 50–51	pages 52–53
☆ 19	☆ 20	☆ 21	☆ 22	☆ 23	☆ 24	☆ 25

pages 54–55	pages 56–57
☆ 26	☆ 27

Did you find all 26 mice?

(Including this one!)